United States General Accounting Office

Report to Congressional Requesters

May 2004

TRANSPORTATION PLANNING

State and Metropolitan Planning Agencies Report Using Varied Methods to Consider Ecosystem Conservation

GAO-04-536

Highlights

Highlights of GAO-04-536, a report to congressional requesters

May 2004

TRANSPORTATION PLANNING

State and Metropolitan Planning Agencies Report Using Varied Methods to Consider Ecosystem Conservation

Why GAO Did This Study

The nation's roads, highways, and bridges are essential to mobility but can have negative effects on plants, animals, and the habitats that support them (collectively called ecosystems in this report). Federally funded transportation projects progress through three planning phases: long range (20 or more years), short range (3 to 5 years), and early project development, (collectively defined as planning in this report) before undergoing environmental review (which includes assessing air and water quality, ecosystems, and other impacts) required under the National Environmental Policy Act. Federal law requires planners to consider protecting and enhancing the environment in the first two phases, but does not specify how and does not require such consideration in the third phase.

GAO reported on (1) the extent to which transportation planners consider ecosystem conservation in planning, (2) the effects of such consideration, and (3) the factors that encourage or discourage such consideration. GAO contacted 36 planning agencies (24 states and 12 of approximately 380 metropolitan planning organizations), as well as officials in 22 resource agencies that maintain ecological data and administer environmental laws. The Department of Transportation and U.S. Army Corps of Engineers had no comments on a draft of this report. The Department of the Interior generally agreed with the contents of our draft report.

www.gao.gov/cgi-bin/getrpt? GAO-04-536.

To view the full product, including the scope and methodology, click on the link above. For more information, contact Katherine Siggerud, (202) 512-2834, siggerudk@gao.gov.

What GAO Found

Of the 36 transportation planning agencies that GAO contacted, 31 considered ecosystem conservation in transportation planning, using a variety of methods. For example, Colorado conducts studies that incorporate ecosystem issues to guide future transportation decisions, uses advance planning to avoid or reduce impacts, and actively involves stakeholders. New Mexico uses planning studies to identify locations where wildlife are likely to cross highways and design underpasses to allow safe crossings. In the absence of specific requirements, federal agencies encourage ecosystem consideration in planning.

Planners and state resource agency officials most frequently reported reduced ecosystem impacts and improved cost and schedule estimates as positive effects. For example, planners in New York changed a planned five-lane highway to a lower-impact two-lane boulevard after weighing the area's mobility needs and the project's impact on the surrounding habitat. In Massachusetts, resource agency officials said that addressing ecological requirements in planning improved schedule certainty during the federally required environmental review. Furthermore, planners and resource agency officials reported that working together has improved relationships between their agencies, thereby allowing ecosystem concerns to be resolved in a more timely and predictable manner. Officials also listed negative effects, such as higher project costs and more work for resource agencies.

Most Frequently Reported Benefits from Considering Ecosystem Conservation

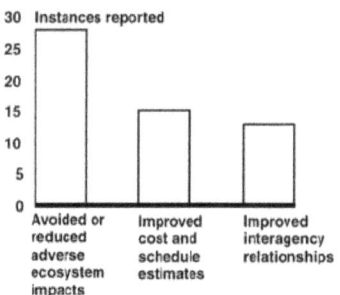

Source: GAO analysis of interview responses.

Constituent support from agency staff, political appointees, or the public was the most frequently reported factor (27 instances) that encouraged planners to consider ecosystem conservation. For example, New Mexico's "pro-environment" culture reportedly encourages planners to consider ecosystem conservation. The cost in time and resources of considering ecosystem conservation was most often cited as a discouraging factor (23 instances). For example, Colorado planners cited the significant amount of time needed to collect and maintain access to ecosystem data.

United States General Accounting Office

Contents

Letter		1
	Results in Brief	3
	Background	7
	Most Planners Contacted Reported Considering Ecosystem Conservation during Transportation Planning	9
	Planners and Resource Agency Officials Reported Mainly Positive Effects of Considering Ecosystem Conservation	20
	Support from Constituents and Transportation Agency Personnel Most Often Encouraged Consideration of Ecosystem Conservation	23
	Agency Comments and Our Evaluation	28
Appendix I	**Telephone Interview Questions for State and Metropolitan Area Planners**	30
	Long-Range Transportation Planning	30
	State Transportation Improvement Program Planning	32
	Pre-NEPA Planning	33
Appendix II	**Telephone Interview Questions for Resource Agency Officials**	35
	State Transportation Planning	35
	Metropolitan Planning Organization Transportation Planning	35
	General Questions	35
Appendix III	**Scope and Methodology**	37
Appendix IV	**Methods Used by Twenty-Two Agencies to Consider Ecosystem Conservation**	43
Appendix V	**Department of the Interior Comments GAO's Mission**	44

Tables

Table 1: Phases During Which Transportation Planning Agencies Consider Ecosystem Conservation — 11
Table 2: Factors that Reportedly Encourage Consideration of Ecosystem Conservation in Transportation Planning — 24
Table 3: Factors that Reportedly Discourage Consideration of Ecosystem Conservation in Transportation Planning — 26

Figures

Figure 1: State and Metropolitan Planning Agencies Surveyed, and Whether They Reported Considering Ecosystem Conservation in Transportation Planning — 5
Figure 2: Reported Consideration of Ecosystem Conservation during Transportation Planning for 36 Planning Agencies — 10
Figure 3: Example of an Underpass Created to Allow Bears to Cross Highway Right-of-way without Danger of Collisions with Vehicles — 14
Figure 4: Effects of Considering Ecosystem Conservation in Transportation Planning Reported by Planners and Resource Agency Officials — 21

Abbreviation

NEPA   National Environmental Policy Act

This is a work of the U.S. government and is not subject to copyright protection in the United States. It may be reproduced and distributed in its entirety without further permission from GAO. However, because this work may contain copyrighted images or other material, permission from the copyright holder may be necessary if you wish to reproduce this material separately.

United States General Accounting Office
Washington, DC 20548

May 17, 2004

Congressional Requesters

The nation's vast network of roads, highways, and bridges is essential to interstate commerce, economic growth, national defense, and leisure mobility. Yet the construction, improvement, rehabilitation, and even maintenance of the tens of thousands of miles of this transportation infrastructure each year can cause permanent environmental change by disturbing plant and animal habitats, creating barriers to animal movement, and producing other impacts. By one estimate, roads ecologically affect about one-fifth of the U.S. land mass.[1]

Although federal agencies must assess the environmental impact of proposed federally funded transportation projects under the National Environmental Policy Act (NEPA), state and metropolitan planners have the opportunity to consider these issues earlier during three planning phases: (1) as they develop long-range (20 or more years) plans; (2) as they develop short-range (3-5 years) plans known as transportation improvement programs; and (3) as they conduct early project planning.[2] The Transportation Equity Act for the 21st Century requires that planners develop these long-range plans and short-range programs and that the plans consider projects and strategies that will, among other things, protect and enhance the environment. However, the act provides no guidance on how planners should meet this requirement.

You requested that we identify the extent to which planners consider the conservation of plants, animals, and the habitats that support them

---

[1] R.T. Forman, "Estimate of the Area Affected Ecologically by the Road System in the United States," *Conservation Biology* (2000) 14(1):31-35, cited in Natasha C. Kline, *The Effects of Roads on Natural Resources: A Primer Prepared for the Sonoran Desert Conservation Plan*, (Tucson, Arizona: January 2002).

[2] Approximately 380 metropolitan planning organizations perform transportation planning for areas having populations of 50,000 or more. State departments of transportation develop and implement statewide transportation plans and generally implement projects listed in metropolitan area plans. The National Environmental Policy Act requires that federal agencies assess the environmental impact of proposed actions that would significantly affect the environment. For a detailed description of how the act affects highway planning, design, and construction, see U.S. General Accounting Office, *Highway Infrastructure: Stakeholders' Views on Time to Conduct Environmental Reviews of Highway Projects*, GAO-03-534 (Washington, D.C.: May 23, 2003).

(collectively called "ecosystems" in this report) in transportation planning.[3] In response, we asked transportation planners and others to identify (1) the extent to which state and metropolitan area transportation planners consider ecosystem conservation and how federal agencies are involved; (2) the effects, if any, of considering ecosystem conservation during transportation planning; and (3) the factors that encourage or discourage transportation planners from considering ecosystem conservation.

To carry out this work, we reviewed laws and regulations relating to transportation planning and ecosystem conservation and spoke with officials of federal transportation agencies, resource agencies (those having responsibility for maintaining ecological data and administering federal environmental laws) and transportation and environmental conservation associations. We also selected a nonprobability sample of 24 states and 12 metropolitan planning organizations, primarily on the basis of geographic diversity, to reflect a variety of ecosystems.[4] We spoke with officials in each of our sample states' departments of transportation and metropolitan planning organizations to ascertain (1) the extent to which, if at all, they consider ecosystem conservation during state and metropolitan area transportation planning before they are required to consider the proposed project's environmental impact under NEPA; (2) anticipated and observed effects of considering ecosystems during transportation planning; and (3) factors that may encourage or discourage planners from considering ecosystems during transportation planning. To gain an understanding of the breadth and depth of each sample states' and metropolitan planning organizations' consideration of ecosystem conservation in transportation planning, we asked a variety of questions about how planners implement this approach, whether and how they involve stakeholders, what types and sources of data they consider, what positive and negative effects they have observed or expect to observe, and what factors encourage and discourage them from these efforts. (See app. I for a complete listing of these questions.) To obtain an additional perspective on the information that planning agencies reported, we contacted officials in resource agencies in 22 of our sample states.[5] We

---

[3]Because federal law already requires that states and local governments meet air and water quality standards, we did not include air and water issues in our review.

[4]A nonprobability sample is a sample not produced by a random process.

[5]We attempted to obtain resource agency perspectives in each of the 24 states in our sample, but were unable to contact two of these agencies.

asked these officials how they are involved in transportation planning, whether they collect ecological data and make these data available to transportation planners, what they believe are the effects of considering ecosystem conservation in transportation planning, and what factors encourage and discourage them from participating in transportation planning. (See app. II for a complete listing of these questions.) Finally, we reviewed transportation plans that were available from the state departments of transportation and metropolitan planning organizations in our sample. Although we requested planners' and resource agency officials' observations about the effects of considering ecosystem conservation in transportation planning, we did not evaluate the effectiveness of their efforts, or determine whether one agency's efforts were more effective than another's. We did not verify the statements of state and metropolitan transportation planners or resource agency officials because it was not practical to do so. The results of our work cannot be projected to all states and metropolitan planning organizations. In order to make reliable generalizations, we would have needed to randomly select a larger sample of states and metropolitan planning organizations than time allowed. We conducted our work from May 2003 through April 2004 in accordance with generally accepted government auditing standards. (See app. III for additional information on our scope and methodology.)

Results in Brief

The majority of the state and metropolitan planners that we contacted reported considering ecosystem conservation in transportation planning, and federal agencies encourage them to do so. (See fig. 1.) Planners in 31 of the 36 agencies (86 percent) described considering ecosystem conservation at varying points in transportation planning using a variety of methods. Planners in four states—Oregon, South Dakota, Colorado, and North Carolina—described extensively considering ecosystem conservation during planning through methods such as studies that incorporate ecosystem issues to guide future transportation decisions, advance planning to avoid or reduce ecosystem impacts, and active stakeholder participation. Twenty-two of the 31 said they conduct corridor studies or use project screening, among other methods, to consider ecosystem conservation.[6] For example, New Mexico used corridor studies

---

[6]A corridor is a broad geographic band that follows a general directional flow connecting major sources of trips that may contain a number of street, highway, and transit route alignments.

to plan for, among other things, where bear and deer were likely to cross highways, and designed underpasses for them at these locations to help prevent vehicle collisions with wildlife. Planners in two agencies described focusing most of their ecosystem conservation efforts on ecological resources within areas of specific interest to their region, such as wetlands. Finally, planners in three agencies reported using mainly resource agency data and input from other stakeholders to determine whether their transportation plans could affect ecosystems, or incorporated in their transportation plans locally developed plans that consider ecosystem conservation. Planners in five agencies said they do not consider ecosystem conservation in transportation planning before projects are subject to federal environmental review because, among other things, these agencies lack the time and resources or guidance on how to do so. Officials we contacted in state wildlife conservation or natural resource departments, as well as similar resource agencies, generally agreed that they assist transportation planners in considering ecosystem conservation during transportation planning. However, 11 of the state resource agency officials said they would like to be more involved in transportation planning or commented that communication with their state departments of transportation could be improved. Although federal law does not specifically require planners to consider ecosystem conservation in transportation plans, the Federal Highway Administration and the U.S. Fish and Wildlife Service encourage state transportation planners to do so. The U.S. Fish and Wildlife Service and the Army Corps of Engineers often assist planners by providing ecosystem data or comments on transportation plans.

Figure 1: State and Metropolitan Planning Agencies Surveyed, and Whether They Reported Considering Ecosystem Conservation in Transportation Planning

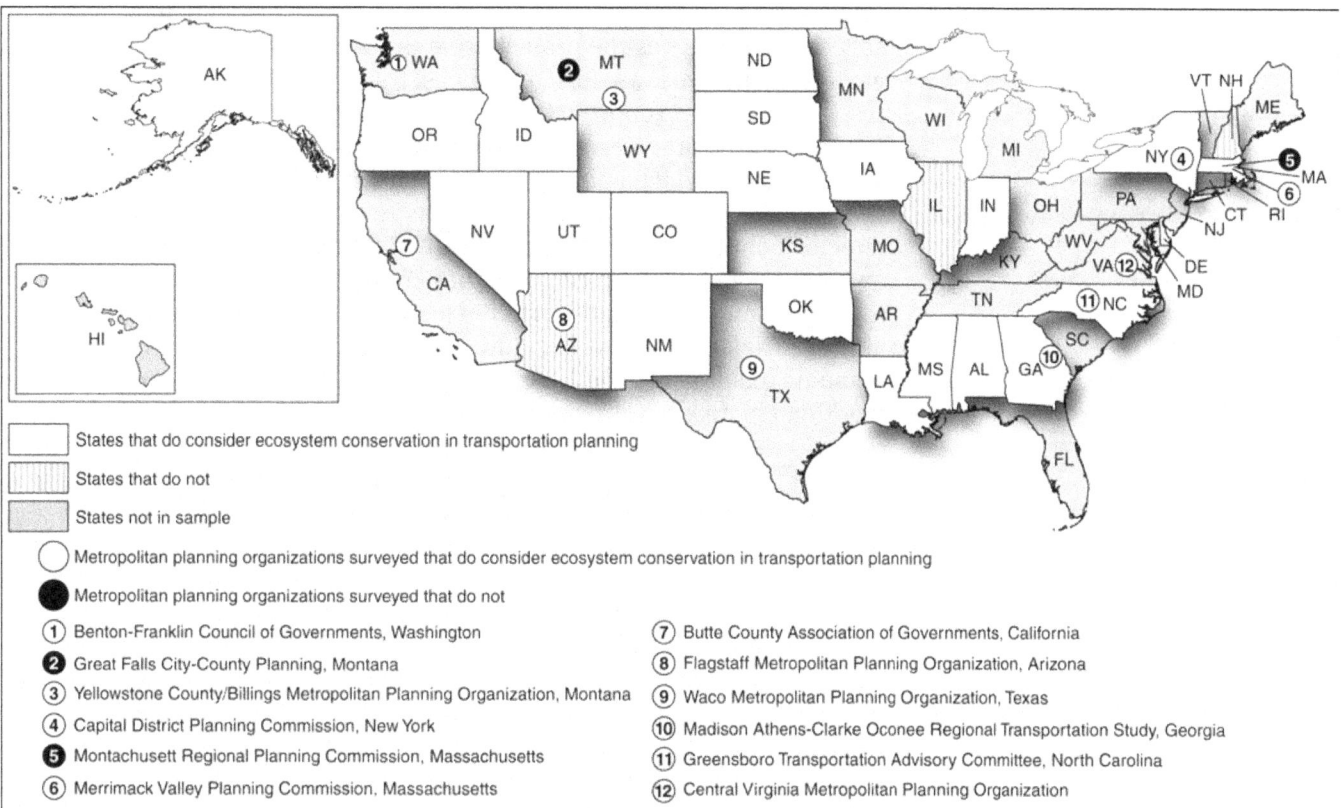

Source: GAO.

The effects of considering ecosystem conservation in transportation planning were mostly positive, according to the planners and state resource agency officials we interviewed. Specifically, planners in 29 of the 31 agencies that consider ecosystem conservation in transportation planning, and 16 of the 19 state resource agency officials that we interviewed in the states that consider ecosystem conservation, described one or more positive effects on the environment. These positive effects include conserving habitat, reducing habitat fragmentation, or scheduling construction times to reduce impacts on breeding of certain species. For example, metropolitan planners in New York told us that they changed plans for a five-lane highway to a lower-impact two-lane boulevard after finding that the wider highway would significantly affect the surrounding habitat and that, according to an updated traffic study, the wider highway was not needed to ensure mobility. In addition, 12 planners and three state resource agency officials reported that considering ecosystem conservation in transportation planning leads to more certain project costs and schedules. In Massachusetts, for example, resource agency officials told us that addressing ecosystem conservation in planning improves schedule certainty as the project progresses through federally required environmental reviews. In 13 instances, transportation planners and state resource agencies reported that working together to address ecosystem issues in transportation planning had improved relationships between their agencies, which allowed environmental issues to be resolved in a timely and predictable manner. Officials also reported other positive effects, including better relationships with the public and a heightened awareness of ecosystem issues among transportation planning staff. On the other hand, eight transportation planners and one state resource agency official reported that addressing ecosystem issues during project planning resulted in negative effects, such as higher project costs and workload increases for resource agencies.

Support from constituents and transportation agency personnel was the key factor that reportedly encouraged transportation planners to consider ecosystem conservation, while the cost in time and resources was the key discouraging factor identified. Of the 31 planners we interviewed who said they considered ecosystem conservation in transportation planning, 27 cited support from staff in their own agencies, political appointees, or the public as an encouraging factor. For example, planners in Oregon and New Mexico told us that the state's pro-environment culture and citizens' concerns about protecting ecological resources encourage them to consider ecosystem conservation. Planners mentioned other encouraging factors that are similar to the positive effects they identified, such as more certain cost estimates and project implementation schedules and fewer adverse effects on ecological resources. The most frequently cited discouraging factor, identified by 23 of the planners we interviewed, was the time and resources required to consider ecosystem conservation. For example, transportation planners in Colorado and North Carolina told us that collecting and maintaining access to the data needed to consider ecosystem conservation in transportation planning, while beneficial, required significant time and resources. The time and resources required were also a factor that discouraged three of the five agencies that do not consider ecosystem conservation in transportation planning. Other planners reported discouraging factors such as difficulty in obtaining stakeholders' involvement and pressure from proponents of development to move forward with projects without considering ecosystem conservation.

The Department of Transportation and U.S. Army Corps of Engineers had no comments on a draft of this report. The Department of the Interior generally agreed with the contents of our draft report and provided technical clarifications, which we incorporated as appropriate.

Background

Federally funded highway projects are typically completed in four phases:

- Planning: State departments of transportation and metropolitan planning organizations begin with a vision and a set of long-term goals for their future transportation system, and translate these into long-range transportation plans and short-range plans known as transportation improvement programs. Although not required by federal law, a state department of transportation may perform additional planning once a project is started, such as consulting with resource agencies to determine the project's potential ecosystem impacts. We refer to this final phase of planning as "pre-NEPA planning" in this report.

- Preliminary design and environmental review: State departments of transportation identify a project's cost, level of service, and construction location; assess the potential effects on environmental resources as required by NEPA; and select the preferred alternative.

- Final design and right-of-way acquisition: State departments of transportation finalize design plans, acquire property, and relocate utilities.

- Construction: State departments of transportation award construction contracts, oversee construction, and accept the completed project.

The Transportation Equity Act for the 21st Century lays out general requirements for transportation planning and consideration of the environment. The act requires that state and metropolitan area long-range plans consider projects and strategies that will, among other things, protect and enhance the environment. It also requires states and metropolitan planning offices to provide the public with an opportunity to comment on the transportation improvement programs. Governors review and approve metropolitan transportation improvement programs within their respective states.

However, the Transportation Equity Act for the 21st Century does not specifically address how ecosystem conservation should be considered in transportation planning. The act does not require that long-range transportation plans contain projects and strategies that protect and enhance the environment, and provides no guidance on how planners are to consider ecosystem conservation. Although the Federal Highway Administration reviews and approves each state's transportation improvement program to, among other things, ensure that the plans meet the requirements of the act, failure to meet these requirements is not reviewable in court.

Congress is considering the 6-year surface transportation reauthorization bill. Separate bills have passed in each chamber.[7] The House bill leaves in place the existing legislation's framework of requiring planners to consider the protection and enhancement of the environment in their plans. The Senate bill provides more explicit language on environmental considerations and new consultation requirements for planners. Specifically, it indicates that protecting and enhancing the environment includes "the protection of habitat, water quality, and agricultural and forest land while minimizing invasive species." Additionally, the Senate bill requires that long-range transportation plans include a discussion of (1) the types of potential habitat mitigation activities that may assist in compensating for habitat loss and (2) the areas that may have the greatest potential to restore and maintain habitat types affected by the plan. Further, the bill requires planning agencies to consult with state and local agencies responsible for protecting natural resources.

In addition to meeting the planning requirements of the Transportation Equity Act for the 21st Century and NEPA, planning agencies must adhere to a number of other federal laws pertaining to transportation and the environment before construction can begin on federally funded projects, including:

- *The Endangered Species Act of 1973* is intended to conserve threatened and endangered species and the ecosystems on which they depend. Section 7 of the act requires federal agencies to ensure that projects they authorize, fund, or carry out, including transportation projects, are not likely to jeopardize the continued existence of any threatened or endangered species (including fish, wildlife, and plants) or result in the destruction or adverse modification of designated critical habitat for these species. The U.S. Fish and Wildlife Service and the National Marine Fisheries Service administer and enforce this law.

---

[7]Safe, Accountable, and Efficient Transportation Equity Act of 2004, S.1072, 108th Cong. Title I(E) (Feb. 26, 2004), and Transportation Equity Act: A Legacy for Users, H.R. 3550, 108th Cong. Title VI (Apr. 2, 2004).

- *The Clean Water Act of 1977* is intended to restore and maintain the chemical, physical, and biological integrity of the nation's waters through the prevention and elimination of pollution. Section 404 of the act pertains to wetland development.[8] Under this section, the Army Corps of Engineers provides permits to transportation agencies whose projects affect wetlands. To obtain permits, applicants must first attempt to avoid adverse impacts to wetlands or, if this is not possible, to minimize the impacts to the extent practicable and compensate for any unavoidable impacts through mitigation.

To comply with these and other laws, transportation planners may coordinate with a variety of state and federal agencies. They do so to obtain ecological data, such as information on threatened and endangered species and wetlands; advice on how to address adverse impacts of transportation projects; or both.

Most Planners Contacted Reported Considering Ecosystem Conservation during Transportation Planning

Of the 36 transportation planners we interviewed, a total of 31 (21 out of 24 in state departments of transportation and 10 out of 12 in metropolitan planning organizations) reported using various methods to consider ecosystem conservation during transportation planning. Some of these 31 planning agencies begin considering ecosystem conservation in transportation planning as they develop their long-range plans while others begin considering ecosystems conservation just prior to starting the federally required environmental review under NEPA. Four of these agencies reported using multiple approaches to consider ecosystem conservation, 22 stressed their use of corridor studies or project screening, 2 emphasized their consideration of the ecological resources of specific interest in the surrounding area, and 3 reported using methods similar to other agencies but do not use corridor studies or project screening or focus on specific resources. (See fig. 2.) Planners in 5 agencies said they do not consider ecosystem conservation during transportation planning. In the absence of specific federal requirements to consider ecosystem conservation in transportation planning, federal agencies encourage state and metropolitan area planners to do so and they provide technical assistance.

---

[8]Wetlands are generally defined as transitional areas such as swamps, marshes, bogs, and similar areas between open waters and dry land.

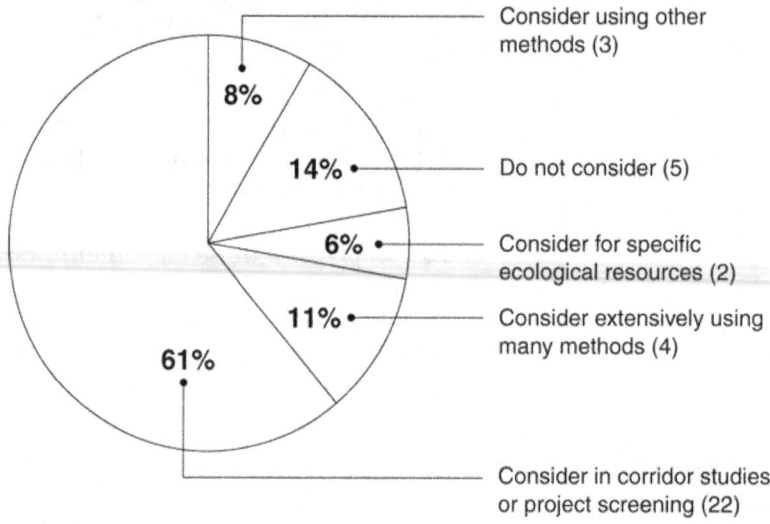

Figure 2: Reported Consideration of Ecosystem Conservation during Transportation Planning for 36 Planning Agencies

Source: GAO analysis of interview responses.

Note: Other methods do not include corridor studies, project screening, or focus on specific ecological resources.

Planning Agencies Vary on How Early They Consider Ecosystem Conservation

Of the 31 planning agencies that consider ecosystem conservation in transportation planning, 21 (68 percent) first do so as they develop their long-range plans. (See table 1.) Four agencies (13 percent) begin considering ecosystem conservation as they develop transportation improvement programs. The remaining six agencies (19 percent) begin just before starting the federally required environmental review under NEPA (pre-NEPA planning). Twenty of 31 agencies reported considering ecosystem conservation at more than one point, and 14 reported considering ecosystem conservation during corridor studies that begin at varying times during planning.

Table 1: Phases During Which Transportation Planning Agencies Consider Ecosystem Conservation

Planning agency	Long-range plan	Transportation improvement program	Pre-NEPA
State planning agencies			
• Alabama			x
• Alaska	x	x	
• Colorado	x		x
• Delaware	x	x	x
• Georgia		x	x
• Idaho	x	x	
• Indiana		x	x
• Iowa			x
• Louisiana	x	x	
• Massachusetts	x		
• Mississippi			x
• Nebraska			x
• Nevada	x	x	
• New Mexico			x
• New York			x
• North Carolina	x	x	x
• North Dakota		x	x
• Oklahoma	x		x
• Oregon	x	x	x
• South Dakota	x	x	x
• Utah		x	
Metropolitan planning organizations			
• Benton-Franklin Council of Governments, Washington	x	x	
• Butte County Association of Governments, California	x		x
• Capital District Transportation Commission, New York	x	x	
• Central Virginia Metropolitan Planning Organization, Virginia	x		x
• Flagstaff Metropolitan Planning Organization, Arizona	x	x	
• Greensboro Transportation Advisory Committee, North Carolina	x		
• Madison Athens-Clarke Oconee Regional Transportation Study, Georgia	x	x	x
• Merrimack Valley Planning Commission, Massachusetts	x	x	
• Waco Metropolitan Planning Organization, Texas	x		
• Yellowstone County/Billings Metropolitan Planning	x		

Source: GAO analysis of interview responses.

Four State Planning Agencies Consider Ecosystem Conservation Using Several Approaches

Oregon, South Dakota, Colorado, and North Carolina reported extensively considering ecosystem conservation in transportation planning using several approaches. The Oregon Department of Transportation has included a policy in its long-range plan to, among other things, maintain or improve the natural and built environment, including fish passage and habitat, wildlife habitat and migration routes, vegetation, and wetlands. The long-range transportation plans of Colorado and North Carolina each contain specific references to goals or policies to conserve ecosystems, while South Dakota's plan contains a less specific goal aimed at protecting the environment.

Oregon planners said they meet monthly with state and federal resource agencies and with the Federal Highway Administration to discuss project proposals before beginning to address NEPA requirements. To plan for each project's potential impact, the planners said they obtain data from a variety of sources, such as field studies led by biologists, the Oregon Natural Heritage Data System, the National Wetlands Inventory, and the state department of transportation's ecological survey of all the roads in the state.[9] The planners then use these data and a set of criteria developed by stakeholders to screen projects before programming them for construction.

The South Dakota Department of Transportation becomes increasingly involved with federal and state resource agency stakeholders—including the U.S. Fish and Wildlife Service; Army Corp of Engineers; U.S. Forest Service; South Dakota Game, Fish, and Parks; and the South Dakota Department of Natural Resources—as a project evolves from a conceptual plan through final design. Initially, the department works with state resource agency stakeholders to obtain ecological data in geographic information system or paper formats that identify ecological resources located within the study boundaries and uses these data to avoid sensitive habitat.[10] The department then develops plans to avoid, minimize, or mitigate the project's impact. Later, when more specific project design plans become available, the department works with resource agency

---

[9]The Oregon Natural Heritage Data System is the state's most comprehensive database of rare, threatened, and endangered species. It includes site-specific information on the occurrences, biology, and status of more than 2,000 species throughout the state.

[10]A geographic information system is a system of computer software, hardware, and data used to manipulate, analyze, and graphically present a potentially wide array of information associated with geographic locations.

stakeholders to determine habitat locations, adjust project alignments to avoid habitat, or consider other design changes to minimize the project's impact before beginning the environmental review required under NEPA.

The Colorado Department of Transportation has assigned one of its employees to work with the U.S. Fish and Wildlife Service to focus on transportation issues, according to state transportation planners. The planners said that numerous stakeholders from federal, state, and nongovernmental agencies assist the department in determining species and habitat locations throughout the state and in focusing efforts on conservation and mitigation planning. The planners reported that the department is conducting advance planning to integrate ecosystem issues into corridor studies that they expect to develop over the life of the long-range plan. They also said that Colorado has established a revolving fund to acquire habitat for mitigation before specific projects are actually developed.

Finally, the North Carolina Department of Transportation considers ecosystem conservation in transportation planning by making extensive use of resource agency personnel and geographic information system data. According to state planners, the department funds 33 resource agency positions to help identify and resolve ecosystem issues early in project development. The planners told us they use the geographic information system data to identify where ecosystems may conflict with transportation plans and determine the potential cost of addressing the conflicts. They said that the department, in partnership with the Army Corps of Engineers, also identifies and acquires property for future mitigation. Finally, the planners said that the department assists small metropolitan planning organizations and localities in broad-based ecosystem screening on all projects to identify any ecological issues and potential costs associated with those issues.

Most Planners Said They Considered Ecosystem Conservation in Transportation Planning When Conducting Corridor Studies or Screening Projects

Twenty-two of the 31 planners who consider ecosystem conservation during transportation planning conduct corridor studies or screen projects for ecosystem impact. These planners survey ecosystems in the corridor and take steps to avoid or mitigate ecological impacts. For example, planners in New Mexico, with data from their Department of Game and Fish, used corridor studies to identify areas of high potential for animal-vehicle crashes. Planners described how such planning studies led to the construction of underpasses that allow bear and deer to pass beneath highways in the state. (See fig. 3.) Nebraska reviews ecological databases to identify potential impacts of planned transportation projects; considers avoidance strategies; and, if avoidance is not possible, documents the conflict so that project designers can develop mitigation measures, according to state transportation planners.

Figure 3: Example of an Underpass Created to Allow Bears to Cross Highway Right-of-way without Danger of Collisions with Vehicles

Source: Florida Department of Transportation.

Some planning agencies screen out projects from their plans that would have undesirable ecosystem impacts. For example, metropolitan planners for the Merrimack Valley area in Massachusetts told us that they use data from a geographic information system in planning to identify ecological resources in the path of proposed projects. Using this information, together with public comments on the project, they determine whether the ecological impacts require that the project be redesigned or terminated prior to beginning the environmental review required under NEPA.

Nearly all planning agencies that develop corridor studies or use ecosystem impacts to screen projects involve stakeholders in developing their plans. For example, Alaska invites federal agencies—including the U.S. Fish and Wildlife Service, Army Corps of Engineers, National Park Service, Bureau of Land Management, and National Marine Fisheries Service—and its Departments of Fish and Game, and Natural Resources to meetings to provide input for transportation plans. After a meeting, each agency has the opportunity to write a letter of concern about specific resources or areas. Metropolitan planning organizations, local governments, municipal officials, tribes, elected officials, and anyone else who has expressed interest in Alaska's transportation planning are also invited to review and comment on transportation plans.

Of the 22 planning agencies that consider ecosystems by conducting corridor studies or project screening, 12 include ecosystem conservation as a policy or goal in their long-range transportation plans. For example, the Central Virginia Metropolitan Planning Organization's long-range transportation plan calls for an assessment of the social and environmental impacts of the transportation plan's recommendations, and establishes the policy of removing projects with unacceptably high environmental or community impacts from planning consideration.

In addition to considering ecosystem conservation in transportation planning through corridor studies or as a means to screen potential projects, these 22 planning agencies reported using one or more of the following common methods either in addition to or in combination with corridor studies or screening:

- using resource agencies as stakeholders in developing transportation plans;

- considering the views of environmental interest groups in developing transportation plans;

- using resource agency data to determine mitigation requirements, develop alternative locations, or to avoid planning projects with unacceptably high ecosystem impact;

- using geographic information systems to determine ecological resource locations;

- providing funding for ecological impact studies;

- having planning agency or resource agency personnel conduct site visits to determine or confirm the location of ecological resources; and

- incorporating in transportation plans local plans that have considered ecosystem conservation.[11]

Six of these agencies reported using at least 4 of the methods listed above. The remaining 16 used 3 or fewer methods. Because we did not evaluate the effectiveness of these methods, the number of methods used by a planning agency does not necessarily indicate effectiveness. (See table 4 in app. IV for a summary of the specific methods that each agency reported using.)

Two Agencies Focus on One or More Specific Ecological Concerns in the Area

Transportation planners in Georgia told us they focus on preserving the state's wetlands through mitigation banking.[12] The state department of transportation has established funding accounts to purchase land for wetland mitigation banking and to pay for consultants to design wetland mitigation banks, according to planners in Georgia. They told us that the department has also entered into a memorandum of agreement with a state resource agency for the long-term maintenance of these mitigation banks. These planners said that nongovernmental organizations, including The Nature Conservancy, Georgia Trust for Public Land, and Georgia Conservancy, help identify properties for sale and conduct on-site reviews of potential sites for wetland mitigation banks. Federal resource agencies assist by reviewing proposed land acquisitions to determine if the land is suitable for use as a wetland mitigation bank, according to the planners. They added that, when transportation projects are at the conceptual design stage, state resource agencies identify wetlands, streams, and endangered species habitats that could be adversely affected by the project and point out avoidance or mitigation opportunities.

Planners in Montana's Yellowstone County/Billings metropolitan area told us that their focus is on the natural resources of the Yellowstone River corridor and the Rim Rocks. These planners said they consider ecosystem conservation in planning transportation projects that would affect these natural resources, primarily through consultations with stakeholders such as the Yellowstone River Parks Association, Bike Net, local government

---

[11] This list includes only those methods reported by at least 2 of the 22 agencies.

[12] A mitigation bank is a site where wetlands, other aquatic resources, or both are restored, created, enhanced, or, in exceptional circumstances, preserved expressly for the purpose of providing compensatory mitigation in advance of authorized impacts to similar resources.

representatives, planning boards, and neighborhood task forces. The planners said these planning boards and neighborhood task forces are involved throughout transportation planning.

Three Agencies Consider Ecosystem Conservation in Transportation Planning Through Other Methods

The Delaware Department of Transportation, Butte County Association of Governments, California, and Madison Athens-Clarke Oconee Regional Transportation Study (the metropolitan planning organization in Athens, Georgia) reported considering ecosystem conservation in transportation planning by using some of the same methods used by other agencies but do not use corridor studies, project screening, or focus on a specific ecological resource. Each of these agencies includes ecosystem conservation as a policy or goal in its long-range transportation plan. Delaware Department of Transportation planners said they consider input from resource agencies and environmental interest groups and use geographic information system data to determine transportation projects' potential impact on ecological resources and develop alternatives as needed. Planners at the Butte County Association of Governments told us they receive input from resource agencies to determine mitigation requirements and use geographic information system data to determine ecological resource locations. Finally, the Madison Athens-Clarke Oconee Regional Transportation Study planners said that local land use plans consider ecosystem conservation as it relates to transportation and they incorporate the local plans in the metropolitan area's transportation plans.

Five Agencies Do Not Consider Ecosystem Conservation in Transportation Planning

Planners in the Arizona, New Hampshire, and Illinois departments of transportation, as well as metropolitan planners in Great Falls City, Montana, and Montachussett, Massachusetts, said they do not consider ecosystem conservation in transportation planning and instead rely on compliance with NEPA to address ecological issues. The reported factors that discouraged these agencies from considering ecosystem conservation in transportation planning include a lack of time and resources required or guidance on how to do so. These factors are discussed in more detail in the final section of this report.

State Resource Agency Officials Generally Agreed That They Are Involved in Transportation Planning but Would Like More Involvement

Resource agency officials in 19 of the 21 states that consider ecosystem conservation in transportation planning generally agreed that they assist transportation planners in doing so. (We were not able to contact resource agency officials in the two remaining states.[13]) However, over half (11) of these resource agency officials said that they would like to be more involved in transportation planning or that communication with their state's department of transportation could be improved. For example, officials of the Oklahoma Department of Wildlife Conservation explained that they need to be involved early in transportation planning because the pressure from supporters of transportation projects often results in concerns about ecosystems surfacing as afterthoughts. Similarly, officials in Utah's Division of Wildlife Resources said that they are involved too late in planning because the project design is already set and budgeting for necessary mitigation sometimes has been inadequate.

Federal Agencies Encourage Consideration of Ecosystem Conservation in Transportation Planning

Although federal law does not specifically require planners to consider ecosystem conservation in transportation plans, the Federal Highway Administration encourages state and metropolitan planners to do so by identifying and promoting exemplary initiatives that are unique, innovative, attain a high-level environmental standard, or are recognized as particularly valuable from an environmental perspective, according to the agency's fiscal year 2004 performance plan. These could be planning or project-level initiatives that involve, for example, designing mitigation projects that support wildlife movement and habitat connectivity, developing watershed-based environmental assessment and mitigation approaches, or using wetland banking. The agency has identified eight such initiatives and plans to identify and promote at least 30 initiatives by September 30, 2007.

North Carolina's Ecosystem Enhancement Program is one of the eight exemplary initiatives that the Federal Highway Administration has identified. In view of a rapidly expanding transportation program with a high volume of projects affecting an estimated 5,000 acres of wetlands and 900,000 feet of streams over 7 years, North Carolina plans to consider and mitigate the potential impacts of many planned projects in a comprehensive manner by assessing, restoring, enhancing, and preserving

---

[13]We asked transportation planners in each state in our sample to provide the name of the resource agency official that they most frequently contacted. Nevada planners did not provide a resource agency contact. We were unable to arrange an interview with the New York resource agency contact.

ecosystem functions and compensating for impacts at the watershed level. This approach to ecosystem conservation aims to decouple ecosystem mitigation from individual project reviews.

Federal Highway Administration officials believe that such integrated approaches help break down organizational barriers between state departments of transportation and state resource agencies. They added that publicizing exemplary initiatives helps show that addressing ecosystem conservation in transportation planning improves working relationships between these agencies and facilitates interagency cooperation in the future. As noted in the next section of this report, many planners and resource agency officials that we interviewed cited improved interagency relationships as a positive effect of considering ecosystem conservation in transportation planning.

The U.S. Fish and Wildlife Service also encourages state departments of transportation and state resource agencies to share project planning and ecosystem information to incorporate more forethought to wildlife habitats, before project designs are set and while flexibility still exists, according to agency officials. To this end, the Service, in cooperation with the International Association of Fish and Wildlife Agencies, has conducted several regional workshops on state wildlife conservation plans. Officials told us that during these workshops they discussed how the plans could be used to provide transportation planners with important information that they could consider in transportation planning.[14]

The U.S. Fish and Wildlife Service and other federal resource agencies also administer and enforce environmental laws and generally help state planners consider ecosystem conservation by responding to requests for data and providing comments on transportation plans. The federal agencies most frequently consulted by the transportation planners we interviewed were the Fish and Wildlife Service and the Army Corps of Engineers. Transportation planners said they often ask these resource agencies to provide ecological data from geographic information systems or ecological maps to help identify and evaluate a project's impact. Many planners also said these federal resource agencies provide technical expertise or actively participate in transportation planning. For example, a

---

[14]States are required to submit, or commit to develop, wildlife conservation plans by October 1, 2005, to be eligible for wildlife conservation grants under the State Wildlife Grant Program. According to U.S. Fish and Wildlife Service officials, all states have committed to develop these plans.

New York Department of Transportation planner told us that the Fish and Wildlife Service and the Army Corps of Engineers provide technical expertise on the long-term impacts of transportation projects on ecosystems.

Planners and Resource Agency Officials Reported Mainly Positive Effects of Considering Ecosystem Conservation

Regardless of the ways planning agencies consider ecosystem conservation in transportation planning, 29 of the 31 transportation planners and 16 of 19 resource agency officials we interviewed reported one or more positive effects of doing so.[15] These officials listed fewer negative effects.

Twenty-eight planners and resource agency officials reported that considering ecosystem conservation in transportation planning enabled them to avoid or reduce adverse impacts on ecological resources—the most frequently reported positive effect. (See fig. 4.) For example, planners and state resource agency officials reported

- preventing irreparable habitat damage in New York by changing planning from a five-lane highway to planning for a lower-impact two-lane boulevard after a study revealed that the original project would be detrimental to the surrounding habitat, and updated traffic studies indicated that the wider highway was not needed to ensure mobility;

- decreasing habitat fragmentation in North Carolina by using geographic information system data on state ecological resources during project planning to avoid or mitigate unacceptable potential impacts on habitat; and

- working with the state resource agency in Nebraska to identify preferred times for construction in order to reduce impacts on the breeding of certain species.

---

[15]This section reflects the results of 50 interviews out of a possible 60 (24 state and 12 metropolitan planners, and 24 state resource agency officials). It does not include any views on ecosystem conservation that may have been expressed by the transportation planners and resource agency officials that we interviewed in the three states that do not consider ecosystem conservation, nor does it include the views of transportation planners in the two metropolitan planning organizations that do not consider ecosystem conservation in transportation planning. Finally, we did not interview resource agency officials in Nevada and New York for reasons previously stated.

Figure 4: Effects of Considering Ecosystem Conservation in Transportation Planning Reported by Planners and Resource Agency Officials

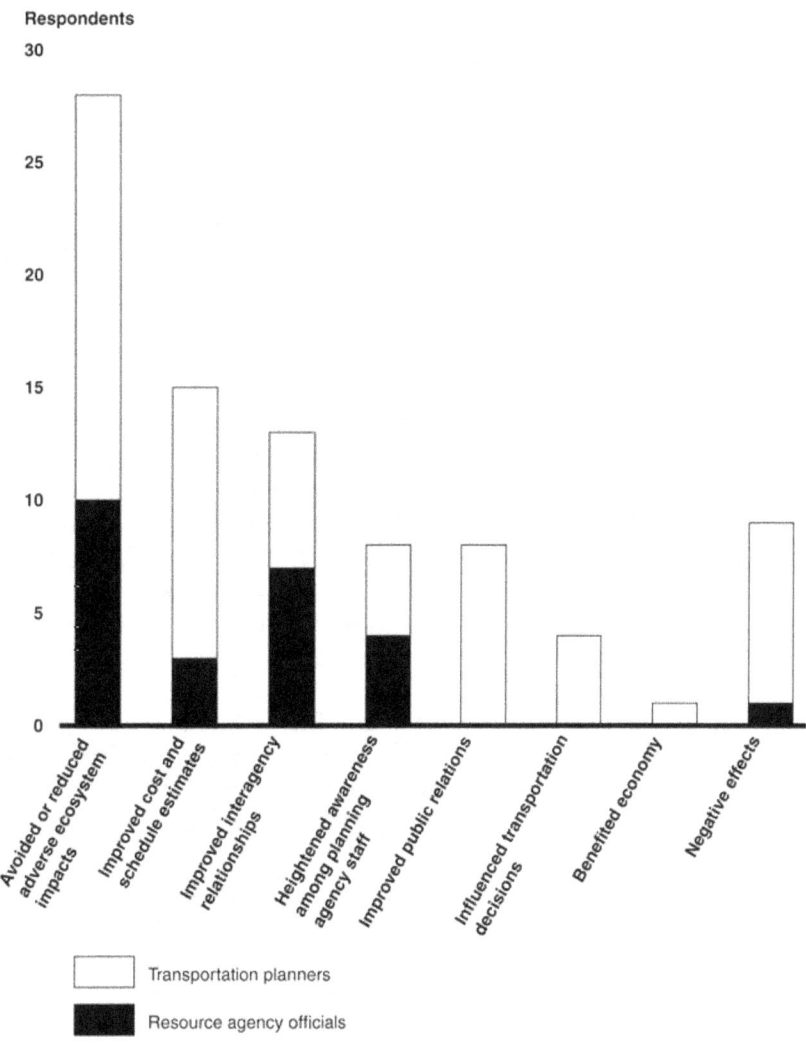

Source: GAO analysis of interview responses.

Fifteen transportation planners and state resource agency officials reported that considering ecosystem conservation improves a project's cost and schedule estimates. For example, planners and state resource agencies reported

- better project cost estimating in Colorado because planners become aware of, and can plan to avoid, unacceptable adverse impacts on ecological resources;

- improved schedule certainty in Massachusetts, because addressing state resource agency requirements during planning provides more certainty that projects will not need to be redesigned to meet these requirements later, during federally required environmental reviews; and

- improved preparedness to address ecological issues during the development of a project in California by identifying those issues early in planning.

In 13 instances, transportation planners and state resource agency officials reported improved relationships between departments of transportation and state resource agencies. For example, improved relationships through partnership and coordination among stakeholders can help resolve environmental issues in a timely and predictable manner. Additional positive effects that planners and state resource agency officials cited include an increased awareness of ecosystem conservation among the transportation planning agency's staff, an improved public image of the department of transportation, and a stimulus to consider transportation alternatives such as transit.

Compared with the number of positive effects attributed to considering ecosystem conservation in transportation planning, planners and resource agency officials reported relatively few negative effects. Planners in South Dakota and at the Benton-Franklin Council of Governments, Washington, told us that considering ecosystem conservation in transportation planning requires additional cost and time. A resource agency official in Iowa said that working with planners to determine project impacts and select mitigation sites adds to the agency's workload. Finally, planners in Louisiana noted that the general public, as well as elected officials who support specific projects, become dissatisfied with the state department of transportation when environmental issues affect a project's delivery.

Support from Constituents and Transportation Agency Personnel Most Often Encouraged Consideration of Ecosystem Conservation

Support from constituents and transportation agency personnel was the most often reported factor that encouraged planners to consider ecosystem conservation in transportation planning. The cost in staff time and money was the most often reported discouraging factor for agencies that reported considering ecosystem conservation. Planners at three of the five agencies, who said they do not consider ecosystem conservation in transportation planning, also cited the cost in time and resources, while the remaining two listed other discouraging factors.

Planners Identified Support from Constituents and Transportation Agency Personnel and Other Encouraging Factors

Twenty-seven of the 31 transportation planners we interviewed, who said they consider ecosystem conservation in transportation planning, cited support from within their own agencies, from political appointees, or from external constituents as a factor that motivated them to do so. (See table 2.) For example, transportation planners in Mississippi told us that their agency is committed to being environmentally aware, and that this culture has encouraged them to consider ecosystem conservation in planning. Metropolitan planners in Albany, New York, noted that their corporate culture provides a strong foundation to consider ecosystem conservation as they develop transportation plans. Similarly, metropolitan planners in central Virginia said that the planning commission's staff are concerned about being good stewards and maintaining a balance between transportation and other concerns.

Table 2: Factors that Reportedly Encourage Consideration of Ecosystem Conservation in Transportation Planning

Encouraging factor	Number of planners reporting
Constituent support and support from transportation agency personnel	27
More certain cost estimates/schedules for project implementation	18
Fewer adverse impacts on ecological resources	7
Improvement in the public's perception of the transportation agency	6
Improved relations with resource agencies	5
Other (each was mentioned only once)	4

Source: GAO analysis of interview responses.

Note: We asked planners to list the three most important factors. This table includes responses from planners in the 31 agencies that consider ecosystem conservation during transportation planning.

The views of elected officials and agency heads were another facet of constituent support. For example, the governor of New York has strongly encouraged planners there to improve their environmental performance, and the governor of New Mexico has initiated a new program that explores several environmental issues, according to planners in those states. This support from elected officials has influenced planners in these states to consider ecosystem conservation during transportation planning. Finally, planners in Delaware and Oregon emphasized the importance of their agency leaders' support for ecosystem conservation.

In addition, the general public's attitude toward ecosystem conservation motivated planners to consider ecosystem conservation during transportation planning. Planners in Oregon and New Mexico attributed their consideration of ecosystem conservation partly to the pro-environment culture in their states. They told us, for example, that citizens are concerned about wildlife protection and view the natural environment as a major asset to the state. Metropolitan planners in Albany, New York, told us that citizens are concerned about excessive land consumption which is one factor that encourages them to consider ecosystem conservation during transportation planning.

Transportation planners also listed encouraging factors that were similar to the positive effects that were discussed earlier in this report. For example, 18 planners said that they were encouraged to consider ecosystem conservation in transportation planning by expectations of more certain cost estimates and construction schedules. Nine of these planners also listed positive effects that centered on developing more accurate cost estimates and determining more predictable project delivery dates. Similarly, seven planners listed having fewer adverse effects on ecological resources as an encouraging factor, while five of these planners also listed this as a positive effect. Planners also listed improved relationships with the state resource agencies as an encouraging factor as well as a positive effect of considering ecosystem conservation in transportation planning.

Planners Identified Time and Resource Requirements and Other Discouraging Factors

Although most of the planners we interviewed reported that considering ecosystem conservation in transportation planning was beneficial, doing so presented challenges. Chief among these challenges was the staff time and money required to consider ecosystem conservation in transportation planning, reported by 23 planners, including those in Arizona, New Hampshire, and Montachusett, Massachusetts, who do not consider ecosystem conservation in transportation planning. (See table 3.) An Arizona planner said that state reductions in funding and staffing have discouraged the department from considering ecosystem conservation during transportation planning, adding that the planning department staff has been reduced by 75 percent since the mid-1990s. New Hampshire planners said they do not have sufficient funds to enter into long-range studies. Therefore, there is pressure to wait until NEPA, which requires, among other things, an assessment of the impact of proposed transportation projects on the natural and human environment.

Table 3: Factors that Reportedly Discourage Consideration of Ecosystem Conservation in Transportation Planning

Discouraging factor	Number of planners reporting
Time and monetary/staffing resources required	23
Difficulty obtaining stakeholder involvement/guidance	15
Political/proponent pressure to move ahead/lack of political support	11
Inappropriate to do so during long-range planning	9
Negative public response/public expectations	6
It is not required	2
Other (each was mentioned only once)	6

Source: GAO analysis of interview responses.

Note: We asked planners to list the three most important factors. This table includes responses from planners in all 36 agencies that we contacted.

The staff time and money required was also the major discouraging factor for those planning agencies that do consider ecosystem conservation in transportation planning. For example, planners in Colorado and North Carolina told us that, while beneficial, it takes a significant amount of time and effort to develop, maintain, and provide access to the data required to consider ecosystem conservation during transportation planning. Additionally, some metropolitan area planners told us that small planning agencies are particularly hard-pressed, because of their small size, to consider ecosystem conservation. For example, a metropolitan planner in central Virginia noted that the limited funding his agency receives for long-range transportation planning precludes more focused activities to address environmental factors, even though the agency would like to do so. Similarly, metropolitan area planners in Athens, Georgia, told us their ability to conduct detailed ecological analyses during planning is very limited because they do not have enough staff.

Difficulties in obtaining involvement or guidance from stakeholders was the second most often cited discouraging factor, according to the planners we interviewed. This was the chief discouraging factor mentioned by a planner in Montachusett, Massachusetts, a metropolitan planning organization that does not consider ecosystem conservation before project developers prepare environmental impact assessments under NEPA. The planner stated that the planning organization lacks guidance from the state or federal agencies on the priority of ecosystem conservation. The planner noted that the planning organization addresses all federal requirements in transportation planning, as well as those issues the state emphasizes, but ecosystem consideration has not been one of them. Planners in Utah, a state that does consider ecosystem conservation in transportation planning, told us that resource agencies prefer to comment on projects that are better defined than is typically the case when they appear in transportation planning documents. On the other hand, a Utah resource agency official told us that his agency would like to be involved in these earlier planning phases, but the state department of transportation does not notify it early enough in planning.[16]

In addition, some planners told us that they lacked guidance from stakeholders, namely state resource agencies, on how to consider ecosystem conservation in transportation planning. They noted that long-term or comprehensive plans for managing the state's ecological resources would help them make decisions about what resources to consider during planning; however, their state resource agencies had not completed such plans. A few of the state and federal resource agencies we interviewed noted, though, that some states are developing wildlife conservation plans as part of a new federal program or other habitat management plans that they believe will be useful to state departments of transportation.

Third, pressure from political leaders or project proponents to move forward in spite of ecological concerns, or because of competing priorities, also discouraged planners from considering ecosystem conservation in transportation planning. For example, planners in North Carolina told us that developers give little credence to environmental concerns. Economic development in Iowa takes precedence over ecosystem concerns, according to a planner there. A state resource agency official in Oregon echoed these sentiments, stating that, in some instances, regional transportation planners and the state department of transportation value improving economic development over conserving ecological resources.

A few other planners cited additional discouraging factors. Local expectations that a project will be built, regardless of ecosystem concerns, is a discouraging factor, according to a transportation planner in North Carolina. Also, planners in three jurisdictions noted that circumstances might change between early planning for a project and its implementation.

---

[16]We did not attempt to reconcile the differences between the statements of Utah planners and resource agency officials.

This was the chief discouraging factor for Illinois, where planners do not consider ecosystem conservation before NEPA. Finally, planners in Great Falls City-County, Montana, a jurisdiction that does not consider ecosystem conservation in transportation planning, stated that their existing policy is to rely on NEPA to assess the ecosystem and other environmental impacts of proposed transportation projects.

Agency Comments and Our Evaluation

The Department of Transportation and U.S. Army Corps of Engineers had no comments on a draft of this report. The Department of the Interior generally agreed with the information in a draft copy of this report and provided technical clarifications, which we incorporated as appropriate. See appendix V for a copy of the Department of Interior's comments.

As arranged with your office, unless you publicly announce its contents earlier, we plan no further distribution of this report until 30 days after the date of this letter. At that time we will send copies of this report to congressional committees with responsibilities for highway and environmental issues; the Secretary of Transportation; the Secretary of the Interior; the Administrator, Federal Highway Administration; the Director, U.S. Fish and Wildlife Service; the Commander, U.S. Army Corps of Engineers; and the Director, Office of Management and Budget. We will also make copies available to others upon request. This report will be available at no charge on our home page at http://www.gao.gov.

If you or your staff have any questions about this report, please contact either James Ratzenberger at ratzenbergerj@gao.gov or me at siggerudk@gao.gov. Alternatively, we may be reached at (202) 512-2834. Key contributors to this report were Jaelith Hall-Rivera, Rebecca Hooper, Jessica Lucas-Judy, Edmond Menoche, James Ratzenberger, and Michelle K. Treistman.

Katherine Siggerud
Director, Physical Infrastructure

List of Congressional Requesters

The Honorable Thomas R. Carper
United States Senate

The Honorable Jon S. Corzine
United States Senate

The Honorable John F. Kerry
United States Senate

The Honorable Carl Levin
United States Senate

The Honorable Ron Wyden
United States Senate

The Honorable Earl Blumenauer
House of Representatives

The Honorable Rosa L. DeLauro
House of Representatives

The Honorable Wayne T. Gilchrest
House of Representatives

The Honorable James L. Oberstar
House of Representatives

Appendix I: Telephone Interview Questions for State and Metropolitan Area Planners

Before each telephone interview with officials at state departments of transportation and metropolitan planning organizations, we provided participants with the following questions and encouraged them to review the questions and to invite others as appropriate to participate in the interview in order to provide as accurate and complete answers as possible. Question numbers preceded by "SLR" are those referring to the development of the long-range transportation plan. Questions preceded by "ST" are those referring to the development of the state transportation improvement program. Finally, questions preceded by "SPN" refer to a phase of project planning that immediately precedes NEPA, which we termed "pre-NEPA planning." Questions for metropolitan area planners were similarly numbered except that they began with the letter "M" to easily differentiate between the state and metropolitan planners' questions and responses.[1]

1) Please answer a, b and c, and follow the instructions as applicable.

   a) Does your state consider ecosystem conservation during the creation of the long-range transportation plan? Yes or No. If yes, answer all SLR questions. If no, answer SLR 7 and SLR 8. In either case, please also answer b and c below.

   b) Does your state consider ecosystem conservation during the creation of the state transportation improvement program? Yes or No. If yes, answer all ST questions. If no, answer only ST 8 and ST 9. In either case, please also answer a and c.

   c) Does your state consider ecosystem conservation during the pre-NEPA phase, or at any other time other than during and after NEPA? Yes or No. If yes, answer all SPN questions. If no, answer only SPN 7 and SPN 8. In either case, please also answer a and b.

Long-Range Transportation Planning

(Answer if applicable.)

SLR1) How does your state consider ecosystem conservation during the creation of the long-range transportation plan?

---

[1]Questions asked of metropolitan area planners were identical except where noted.

Appendix I: Telephone Interview Questions
for State and Metropolitan Area Planners

SLR2) What stakeholders, if any, are involved in helping you consider ecosystem conservation in the long-range transportation plan (federal or state agencies, non-government organizations, other)?

SLR3) How are these stakeholders involved in helping you consider ecosystem conservation in the long-range transportation plan?

SLR4) What type of ecosystem data, if any, do you include in the development of the long-range transportation plan?

SLR5) Please provide any other ways, not discussed above, that your state considers ecosystem conservation when developing the long-range transportation plan.

We would now like to discuss the effects of considering ecosystem conservation in developing the long-range transportation plan.

SLR6) Please describe any anticipated or observed effects, positive or negative, that you can attribute to the consideration of ecosystem conservation in the long-range transportation plan.

We would like to know about factors that encourage or discourage consideration of ecosystem conservation in long-range transportation planning.

SLR7) Please list the three factors that have been the most important in encouraging your state to consider ecosystem conservation as the long-range transportation plan is developed.

SLR8) Similarly, please list the three factors that have been the most important in discouraging your state to consider ecosystem conservation as the long-range transportation plan is developed.

Appendix I: Telephone Interview Questions
for State and Metropolitan Area Planners

State Transportation Improvement Program Planning

We would like to learn about how your state considers ecosystem conservation as it develops the state transportation improvement program.[2]

(Answer if applicable)

ST1) How does your state consider ecosystem conservation during the creation of the state transportation improvement program?

ST2) What stakeholders, if any, are involved in helping you consider ecosystem conservation in the state transportation improvement program (federal or state agencies, non-government organizations, other)?

ST3) How are these stakeholders involved in helping you consider ecosystem conservation in the state transportation improvement program?

ST4) What type of ecosystem data, if any, do you include in the development of the state transportation improvement program?

ST5) Do you use project criteria that incorporate ecosystem conservation when determining which projects will be placed on the state transportation improvement program?

ST6) Please provide any other ways, not discussed above, that your state considers ecosystem conservation when developing the state transportation improvement program.

We would now like to discuss the effects of considering ecosystem conservation in developing the state transportation improvement program.

ST7) Please describe any anticipated or observed effects, positive or negative, that you can attribute to the consideration of ecosystem conservation in the state transportation improvement program.

We would like to know about factors that encourage or discourage consideration of ecosystem conservation in the creation of the state transportation improvement program.

---

[2] In the metropolitan planning organization interviews, we asked the same questions but about the transportation improvement program.

Appendix I: Telephone Interview Questions
for State and Metropolitan Area Planners

ST8) Please list the three factors that have been the most important in encouraging your state to consider ecosystem conservation as the state transportation improvement program is developed.

ST9) Similarly, please list the three factors that have been the most important in discouraging your state to consider ecosystem conservation as the state transportation improvement program is developed.

Pre-NEPA Planning

We would like to learn about how your state considers ecosystem conservation as it begins project development—after the project has been listed on the state transportation improvement program, but before the NEPA process begins. As previously discussed, we call this phase the "pre-NEPA" phase.

(Answer if applicable)

SPN1) How does your state consider ecosystem conservation during the pre-NEPA phase?

SPN2) What stakeholders, if any, are involved in helping you consider ecosystem conservation during the pre-NEPA phase (federal or state agencies, non-government organizations, other)?

SPN3) How are these stakeholders involved in helping you consider ecosystem conservation during the pre-NEPA phase?

SPN4) What type of ecosystem data, if any, do you include in the pre-NEPA phase?

SPN5) Please provide any other ways, not discussed above, that your state considers ecosystem conservation in the pre-NEPA phase.

We would now like to discuss the effects of considering ecosystem conservation in the pre-NEPA phase.

SPN6) Please describe any anticipated or observed effects, positive or negative, that you can attribute to the consideration of ecosystem conservation in the pre-NEPA phase.

We would like to know about factors that encourage or discourage consideration of ecosystem conservation in the pre-NEPA phase.

Appendix I: Telephone Interview Questions for State and Metropolitan Area Planners

SPN7) Please list the three factors that have been the most important in encouraging your state to consider ecosystem conservation during the pre-NEPA phase.

SPN8) Similarly, please list the three factors that have been the most important in discouraging your state to consider ecosystem conservation during the pre-NEPA phase.

Is there anything else that you would like to tell us about considering ecosystem conservation in transportation planning?

We would like to contact someone in the state resource agency (Department of Natural Resources, Department of Environmental Protection, etc.) that is most involved with your agency in considering ecosystem conservation during the transportation planning process. Please provide the name, official title, and contact information.

Appendix II: Telephone Interview Questions for Resource Agency Officials

Prior to each interview with officials at state resource agencies, we provided participants with the following questions and encouraged them to review the questions and to invite others as appropriate to participate in the interview in order to provide as accurate and complete answers as possible. "RA" precedes all question numbers so that we could easily distinguish questions and responses as those pertaining to resource agencies.

State Transportation Planning

RA1) The _____ state department of transportation told us that your agency is involved in transportation planning. Please describe your involvement.

RA2) How did your agency become involved in state transportation planning?

Metropolitan Planning Organization Transportation Planning

RA3) Is your agency involved with metropolitan planning organizations in considering ecosystem conservation in the transportation planning process? If yes, please continue. If no, please skip to RA7.

RA4) In what ways is your agency involved with metropolitan planning organizations in considering ecosystem conservation in transportation planning?

RA5) What metropolitan planning organizations are you involved with? (If you do not know the names of the metropolitan planning organizations, simply list the number that you are involved with.)

RA6) How did your agency become involved in metropolitan planning organization transportation planning?

General Questions

RA7) Does your agency collect or generate ecosystem data? Yes or No.

If yes:

Is it available to state departments of transportation?
Is it available to metropolitan planning organizations?

We would now like to discuss the effects of considering ecosystem conservation in any phase of transportation planning.

Appendix II: Telephone Interview Questions for Resource Agency Officials

RA8) Please describe any anticipated or observed effects, positive or negative, that you can attribute to the consideration of ecosystem conservation in transportation planning prior to NEPA.

We would now like to ask you about factors that encourage or discourage your participation in the consideration of ecosystem conservation in transportation planning.

RA9) Please list the three factors that you consider to be the most important in encouraging your agency to participate in consideration of ecosystem conservation in transportation planning.

RA10) Please list the three factors that you consider the most important in discouraging your agency from participating in consideration of ecosystem conservation in transportation planning.

RA11) Is there anything else you would like to tell us about considering ecosystem conservation in transportation planning?

Thank you.

Appendix III: Scope and Methodology

To obtain a basic understanding of how transportation planners consider ecosystem conservation in transportation planning and how federal agencies are involved, we discussed transportation laws, regulations, and planning procedures with officials in the following agencies:

- Federal Highway Administration in headquarters and Phoenix, Arizona; U.S. Fish and Wildlife Service in headquarters, Phoenix and Tucson, Arizona, and Denver, Colorado; and Army Corps of Engineers in headquarters, Baltimore, Maryland, and Phoenix, Arizona.

- State departments of transportation, resource agencies, and metropolitan planning organizations in Virginia, Massachusetts, Wisconsin, Mississippi, and Colorado; the metropolitan planning organizations for the Washington, D.C., area and Pima County, Arizona; and state departments of transportation and resource agencies in Florida and Maryland.

- The American Association of State Highway Transportation Officials, Association of Metropolitan Planning Organizations, The Nature Conservancy, International Association of Fish and Wildlife Agencies, and Defenders of Wildlife.

At each of these locations, we also obtained and reviewed transportation planning documents. We defined ecosystems as plants and animals and the habitats that support them. We defined planning as activities associated with developing the federally required long-range transportation plan, short-range transportation improvement program, and the nonfederally required project planning that some jurisdictions perform just prior to beginning the environmental review required by the National Environmental Policy Act (NEPA), as well as any activities, such as corridor studies, that are performed concurrently with, but independently of, federally mandated transportation planning activities. Because federal law already requires that states and local governments meet air and water quality standards, our inquiry did not include identifying whether state departments of transportation and metropolitan planning organizations were considering these issues in transportation planning.

To identify (1) how state and metropolitan area transportation planners consider ecosystem conservation and how federal agencies are involved, (2) the effects these planners have seen from this consideration, and (3) the factors that encourage or discourage them from doing so, we developed a set of questions to ask transportation planners selected through a nonprobability sample of 24 states and 12 metropolitan planning

Appendix III: Scope and Methodology

organizations. We divided the nation into eight geographic zones containing a roughly equal number of states to ensure that our sample was geographically and ecologically diverse. To ensure that our sample included states with a variety of population sizes, we used census data to divide states in each zone into three subgroups according to population—high, low, and medium. We then randomly selected 1 state from each of the 24 subgroups to obtain a 24-state sample, which included the following states:

- Alabama
- Alaska
- Arizona
- Colorado
- Delaware
- Georgia
- Idaho
- Illinois
- Indiana
- Iowa
- Louisiana
- Massachusetts
- Mississippi
- Nebraska
- Nevada
- New Hampshire
- New Mexico

Appendix III: Scope and Methodology

- New York
- North Carolina
- North Dakota
- Oklahoma
- Oregon
- South Dakota
- Utah

To ensure ecosystem diversity among the 12 metropolitan planning organizations in our sample, we divided the nation into quadrants containing a roughly equal number of states. Then, to ensure that our sample would reflect the varying extent to which metropolitan planning organizations consider ecosystem conservation in transportation planning, we used the results from our 2002 survey of all metropolitan planning organizations. The survey asked how much consideration, if any, they give to the impact of transportation projects on environmentally sensitive lands, such as wetlands, when they develop their transportation plans.[1] According to their answers, we divided the metropolitan planning organizations in each quadrant into three subgroups: (1) those that indicated little or no, or some consideration; (2) those that indicated moderate consideration; and (3) those that indicated great or very great consideration. We then randomly selected one metropolitan planning organization from each of the 12 subgroups, resulting in the following sample:

- Benton-Franklin Council of Governments, Washington;
- Butte County Association of Governments, California;
- Capital District Transportation Commission, New York;
- Central Virginia Metropolitan Planning Organization, Virginia;

---

[1] U.S. General Accounting Office, *Environmental Protection: Federal Incentives Could Help Promote Land Use That Protects Air and Water Quality*, GAO-02-12 (Washington, D.C.: Oct. 31, 2001).

Appendix III: Scope and Methodology

- Flagstaff Metropolitan Planning Organization, Arizona;
- Great Falls City-County Planning, Montana;
- Greensboro Transportation Advisory Committee, North Carolina;
- Madison Athens-Clarke Oconee Regional Transportation Study, Georgia;
- Merrimack Valley Planning Commission, Massachusetts;
- Montachusett Regional Planning Commission, Massachusetts;
- Waco Metropolitan Planning Organization, Texas; and
- Yellowstone County/Billings Metropolitan Planning Organization, Montana.

To gain an understanding of the breadth and depth of each sample state's and metropolitan planning organization's consideration of ecosystem conservation in transportation planning, we developed a variety of questions about how planners implement this consideration, whether and how they involve stakeholders, what types and sources of data they consider, what positive and negative effects they have observed or expect to observe, and what factors encourage and discourage them from these efforts. (See app. I for a complete listing of these questions.) Through telephone interviews, we asked planners to address these questions for each of three phases of transportation planning: (1) as they develop their long-range transportation plans, (2) as they develop their short-range transportation improvement programs, and (3) in the project planning stage that immediately precedes the environmental review under NEPA.[2] Planners reported similar effects of considering ecosystem conservation in transportation, planning and similar encouraging and discouraging factors across these three phases. Therefore, we did not report answers to these questions by phase. Appendix II contains the questions that we asked planners who we interviewed in state departments of transportation and metropolitan planning organizations. We also reviewed the available long-range transportation plans of each state and metropolitan planning organization in our samples to determine whether these plans contained goals related to ecosystem conservation.

---

[2] We asked all planners the same questions. We did not provide the planners with sets of possible responses from which to choose.

Appendix III: Scope and Methodology

To obtain the perspectives of state resource agency officials, we asked officials at each department of transportation in our sample to identify the official at the state resource agency who was most involved with the department of transportation during planning.[3] We conducted telephone interviews with resource agency officials in 22 of our 24 sample states, asking these officials how they participate in considering ecosystem conservation in transportation planning, whether they collect ecological data and make these data available to transportation planners, the effects that they can attribute to considering ecosystem conservation, and the factors that encourage or discourage their participation.[4] See appendix II for a complete listing of the questions that we asked resource agency officials.

In analyzing our interview responses, we used content analysis and consensus agreement among four analysts to categorize similar responses, and grouped state and metropolitan planning organizations accordingly. To increase the reliability of our coding of responses, we used consensus agreement among the same four analysts. We did not verify the accuracy of the information that we obtained in our interviews or determine whether or how the consideration of ecosystem conservation that planners described affected transportation projects or ecosystems because it was not practical to do so. However, the variety of questions that we asked of transportation planners, combined with the perspectives of resource agency officials, mitigates the potential that our results portray more extensive consideration of ecosystem conservation in transportation planning than may actually exist. Although we requested planners' and resource agency officials' observations about the effects of considering ecosystem conservation in transportation planning, we did not evaluate the effectiveness of their efforts, or determine whether one agency's efforts were more effective than another's. The results of our work cannot be projected to all states and metropolitan planning organizations. In order to make reliable generalizations, we would have needed to randomly

---

[3] Because state resource agencies are organized in a variety of ways, independently identifying the appropriate resource agency contact in each of our 24 sample states was not practical.

[4] We asked each resource agency official the same questions. We did not provide these officials with sets of possible responses from which to choose. Nevada did not provide a state resource agency contact, and the New York state resource agency contact did not respond to requests to be interviewed. Because Idaho Department of Transportation officials told us that their primary resource agency contact was with the Boise office of the U.S. Fish and Wildlife Service, we interviewed an official at that agency, rather than a state resource agency official.

Appendix III: Scope and Methodology

select a larger sample of states and metropolitan planning organizations than time allowed.

Appendix IV: Methods Used by Twenty-Two Agencies to Consider Ecosystem Conservation

Method used	State planning agencies															Metropolitan planning organizations						
	Alabama	Alaska	Idaho	Indiana	Iowa	Louisiana	Massachusetts	Mississippi	Nebraska	Nevada	New Mexico	New York	North Dakota	Oklahoma	Utah	Benton-Franklin Council of Governments, Washington	Capital District Transportation Commission, New York	Central Virginia	Flagstaff, Arizona	Greensboro Transportation Advisory Committee, North Carolina	Merrimack Valley Planning Commission, Massachusetts	Waco, Texas
Incorporating in transportation plans the local plans that have considered ecosystem conservation																●	●		●			
Using resource agencies as stakeholders in developing transportation plans	●	●	●		●	●	●		●	●	●	●	●	●	●					●	●	●
Considering input from environmental interest groups in developing transportation plans							●					●							●		●	
Having planning agency or resource agency personnel conduct site visits to determine or confirm the location of ecological resources	●			●				●	●			●										
Using resource agency data to determine mitigation requirements, develop alternative locations, or to avoid planning projects with unacceptably high ecosystem impact		●	●			●			●		●	●		●				●	●			●
Using geographic information systems to determine ecological resource locations				●			●						●							●	●	
Providing funding for ecological impact studies		●												●	●							

Source: GAO analysis of interview responses.

Note: The twenty-two agencies included in this appendix are those that employ corridor studies or screen projects for ecosystem impact. The list of methods used does not include every method used by these agencies. It includes only those methods reported as used by two or more agencies.

Appendix V: Department of the Interior Comments GAO's Mission

United States Department of the Interior
OFFICE OF THE SECRETARY
Washington, DC 20240

APR 2 3 2004

Mr. James Ratzenberger
Assistant Director
U.S. General Accounting Office
441 G Street, N.W.
Washington, D.C. 20548

Dear Mr. Ratzenberger:

Thank you for providing the Department of the Interior the opportunity to review and comment on the draft U.S. General Accounting Office report entitled, *"Highway Infrastructure: Many State and Regional Agencies Report Considering Ecosystem Conservation during Planning,"* GAO-04-536, dated March 26, 2004. In general, we agree with the findings that pertain to the U.S. Fish and Wildlife Service.

The Department is pleased with the finding that most agencies found benefits to considering various aspects of ecosystem conservation in their planning efforts. We believe that benefits to transportation planning will continue to increase as agencies expand their efforts to consider ecosystem conservation. The Department is working cooperatively with other Departments under Executive Order 13274 to streamline the environmental review process and promote environmental stewardship for transportation projects. We believe through the cooperative efforts under this Executive Order, project costs and workload can be more effectively managed while fulfilling our stewardship responsibilities.

The enclosure provides specific comments from the Service. We look forward to receiving the final report.

Sincerely,

Assistant Secretary for Fish
and Wildlife and Parks

Enclosure

Note: We included these specific comments in this final report, where appropriate.

GAO's Mission	The General Accounting Office, the audit, evaluation and investigative arm of Congress, exists to support Congress in meeting its constitutional responsibilities and to help improve the performance and accountability of the federal government for the American people. GAO examines the use of public funds; evaluates federal programs and policies; and provides analyses, recommendations, and other assistance to help Congress make informed oversight, policy, and funding decisions. GAO's commitment to good government is reflected in its core values of accountability, integrity, and reliability.
Obtaining Copies of GAO Reports and Testimony	The fastest and easiest way to obtain copies of GAO documents at no cost is through the Internet. GAO's Web site (www.gao.gov) contains abstracts and full-text files of current reports and testimony and an expanding archive of older products. The Web site features a search engine to help you locate documents using key words and phrases. You can print these documents in their entirety, including charts and other graphics. Each day, GAO issues a list of newly released reports, testimony, and correspondence. GAO posts this list, known as "Today's Reports," on its Web site daily. The list contains links to the full-text document files. To have GAO e-mail this list to you every afternoon, go to www.gao.gov and select "Subscribe to e-mail alerts" under the "Order GAO Products" heading.
Order by Mail or Phone	The first copy of each printed report is free. Additional copies are $2 each. A check or money order should be made out to the Superintendent of Documents. GAO also accepts VISA and Mastercard. Orders for 100 or more copies mailed to a single address are discounted 25 percent. Orders should be sent to: U.S. General Accounting Office 441 G Street NW, Room LM Washington, D.C. 20548 To order by Phone: Voice: (202) 512-6000 　　　　　　　　　　TDD: (202) 512-2537 　　　　　　　　　　Fax: (202) 512-6061
To Report Fraud, Waste, and Abuse in Federal Programs	Contact: Web site: www.gao.gov/fraudnet/fraudnet.htm E-mail: fraudnet@gao.gov Automated answering system: (800) 424-5454 or (202) 512-7470
Public Affairs	Jeff Nelligan, Managing Director, NelliganJ@gao.gov (202) 512-4800 U.S. General Accounting Office, 441 G Street NW, Room 7149 Washington, D.C. 20548